Stefan Luckhaus

Cost Estimation in Agile Software Development

Utilizing Functional Size Measurement Methods

www.tredition.de

© 2016 Stefan Luckhaus

Publisher: tredition GmbH, Hamburg

ISBN
Paperback: 978-3-7345-4371-5
Hardcover: 978-3-7345-4372-2
e-Book: 978-3-7345-4373-9

Printed in Germany

This work, including all of its parts, is protected by copyright. Any utilization without the consent of the publisher and the author is prohibited. This applies in particular to electronic reproduction or any other copying, translating, distributing or making contents publicly available.

Contents

Characteristics and Importance of Agile Software Development ... 9
Genesis ... 9
Status quo ... 12
Reliability ... 14

Direct and indirect Cost Estimation Methods ... 17
Principle of Incremental Development ... 17
Expert Estimations ... 19
Indirect Estimations with Story Points ... 20
Indirect Estimations by measuring the Functional Size ... 21
Summary ... 24

Methods for Functional Size Measurement ... 25
Function Point Analysis ... 25
COSMIC Method ... 30
Data Interaction Point Method ... 33
Extending Methods for measuring the Size of Further Development ... 36
The Impact of Complexity ... 36
The Complexity of an Implementation ... 36
Interactional Complexity ... 37
Algorithmic Complexity ... 38
Method Comparison ... 39
Further Methods ... 40

Measuring the Reference Value for an indirect Cost Estimation 43
Regular Measuring of Productivity 43
Process Scope regarding Work 44
Process Scope regarding Sub-Processes 44
Process Scope regarding Quality 45
Automated Measurements 46
Mapping of Objects to be counted on Structural Characteristics 46
Possible Restrictions 48
Iterative Refinement of the measured Productivity 49
Considering non-functional Requirements 52
Regular Measurements 53
The Relationship of Productivity and Quality 54

Conclusion 57

Glossary 59

Bibliography 63

About the Author 65

Introduction

Indirect estimations of development costs, which put the methodically determined size of planned software into relation with a precisely measured value of the own productivity, are a best practice approach for planning software development projects. However, utilizing them requires a minimum degree of specification. Briefly described user stories must be refined by use cases and elementary processes. As a consequence, the size of the planned software will be rendered measurable, while the measurement process is lean and does not require much effort.

This book describes briefly and based on the author's own experience the basics of methodological cost estimations. It demonstrates that this approach agrees well with agile development and especially supports principles such as

- the flexible consideration of new or changed requirements and
- continuous improvement due to retrospectives.

Characteristics and Importance of Agile Software Development

Genesis

At the beginning of the 90s many large-scale projects ran into difficulties – due to their long process times, rigid roles and inflexible structures in conjunction with frequently changing requirements. In this context the C3 project of the Chrysler group (Chrysler Comprehensive Compensation) is often mentioned, which had applied the waterfall model in the beginning. In these days, many US companies experimented with lightweighted development processes and found out that shorter process times, a closer and self-responsible collaboration of the project teams or the uncomplicated handling of change requests lead to a better mitigation of typical risks and in consequence to more successful projects - successful in terms of early benefits by the customer. Process models such as Scrum or Crystal were developed. The Chrysler C3 project could be prevented from failing by introducing some of these lightweighted methods, which afterwards became popular as Extreme Programming [Wells 2009].

In February 2001, experts exchanged their experience with software development processes at a meeting in Utah (USA) and formulated a system of values, laying the foundation for the way of software development which, since then, is called agile – the Agile Manifesto [Agile Manifesto 2001]:

> *We are uncovering better ways of developing software by doing it and helping others do it. Through this work we have come to value:*
>
> **Individuals and interactions**
> *over processes and tools*
>
> **Working software**
> *over comprehensive documentation*
>
> **Customer collaboration**
> *over contract negotiation*
>
> **Responding to change**
> *over following a plan*
>
> *That is, while there is value in the items on the right, we value the items on the left more.*

The Agile Manifesto is often misinterpreted. It is commonly used as an excuse to forego any documentation. However, especially the last paragraph makes it evident that it is just a matter of priorities, and that activities such as documentation are adjudged as certainly valuable.

This system of values was refined by the following twelve principles [Agile Manifesto Principles 2001]:

1. *Our highest priority is to satisfy the customer through early and continuous delivery of valuable software.*

2. *Welcome changing requirements, even late in development. Agile processes harness change for the customer's competitive advantage.*

3. *Deliver working software frequently, from a couple of weeks to a couple of months, with a preference to the shorter timescale.*

4. *Business people and developers must work together daily throughout the project.*

5. *Build projects around motivated individuals. Give them the environment and support they need, and trust them to get the job done.*

6. *The most efficient and effective method of conveying information to and within a development team is face-to-face conversation.*

7. *Working software is the primary measure of progress.*

8. *Agile processes promote sustainable development. The sponsors, developers, and users should be able to maintain a constant pace indefinitely.*

9. *Continuous attention to technical excellence and good design enhances agility.*

10. *Simplicity--the art of maximizing the amount of work not done--is essential.*

11. *The best architectures, requirements, and designs emerge from self-organizing teams.*

12. *At regular intervals, the team reflects on how to become more effective, then tunes and adjusts its behavior accordingly.*

The success of agile methods got around. According to a study performed by Forrester Research at the end of 2005, 14% of the companies in Europe and North America were already using agile processes in software development, while an additional 19% intended to do so in future [Forrester 2005].

Status quo

The technology provider VersionOne, specialized in agile methods, ascertained in 2013 through its seventh annual survey on agile methods that already 84% of all companies employed agile development processes [VersionOne 2013]. In the meantime this number might have increased even further. Even if hybrid process models are applied instead of pure agile ones such as Scrum, e.g. by combining agile methods with project management standards, the following three core characteristics of agile development often remain:

- **Incremental**: Parts of the system are developed at different points in time. Each completed part complements the system.

- **Learning**: The team learns though retrospectives, while the organization learns through a continuous improvement process, e.g. on basis of steady root cause analyses.
- **Immediate**: All parties cooperate closely and directly. The team is characterized by a flexible distribution of tasks without rigid role assignments and its collective responsibility for the product. Customers respectively product owners engage actively and continuously in the development process.

It can be generally stated that today, the "early and continuous delivery of valuable software" as well as the welcoming of "changing requirements, even late in development", which both belong to the principles of the Agile Manifesto, are crucial factors of success for software developing companies. Reasons for this are the rapid penetration of our modern world with software as well as digitalization and virtualization along with global competition. Today, commercial development projects, which leave the customer waiting for his product for more than half a year and give him no chance to get changed requirements considered once the product is in the works, are almost inconceivable.

Reliability

If software is developed for commercial purpose and not just for fun, it is essential to plan and control the achievement of objectives - especially in case the objectives are set in a contract.

Contractual agreements and agile development are not mutually exclusive. Many principles of the Agile Manifesto are derived from reviews and problem analyses of development projects with heavy processes. Therefore they are good measures to mitigate the typical risks of development projects with binding deadlines. Some examples are listed below - with no claim to completeness:

Risk #1:

Invalid planning of the manpower required to fulfil a development contract within the agreed time.

Measure:

This risk can be countered by applying a cost estimation method with accuracy improving with every retrospective. Such methods are described in detail within the following chapters.

Risk #2:

Objectives and requirements of the customer change, before the development is completed, for example as a reaction to unexpected market changes.

Measures:

One principle of the Agile Manifesto is: "Welcome changing requirements, even late in development. Agile processes harness change for the customer's competitive advantage." Thus, agile software development is generally characterized by the willingness to consider new or changed requirements. However, this does not preclude a consensus between the contract parties regarding the commercial effects of such changes.

Risk #3:

The requirements of the customer are not covered completely. His expectations for example regarding usability have not been fulfilled.

Measures:

The product is developed incrementally, with the cycles being as short as possible. The result of each cycle is operable software, which can be checked by the customer regarding the fulfillment of his expectations. The customer, in particular his business and technical professionals, collaborate closely and intensively with the development team.

Risks #4, #5 and #6:

- Wrong assessment of the development progress
- Delays due to communication problems
- Loss of know-how by staff turnover

Measures:

In agile development processes these risks are mitigated by a high level of teamwork. In short daily team meetings problems can be addressed, the progress can be estimated together, pending tasks can be assigned or even tasks in progress can be re-assigned if it seems reasonable. The avoidance of too rigid role and task assignments makes the team robust toward the loss of individuals.

Risk #7:

Additional expenses due to quality deficiencies which have been discovered too late.

Measures:

One characteristic of agile development are automated processes for the daily build of an executable system, where an initial code analysis is performed, followed by functionality checks of particular components using unit tests and, if possible, the execution of scripted test cases.

Direct and indirect Cost Estimation Methods

Principle of Incremental Development

In incremental software development the set of requirements for a planned product is distributed to several sub processes which always result in a new product increment. In agile process models these sub processes are called sprints and they have durations between 2 and 4 weeks. It is recommended to plan a bit more time for the initial sprint – for example to stabilize the architecture, to create a mostly complete data model and also considering that the learning curve of the team is still flat.

In agile development, the expression story respectively user story has been established for brief requirements. These are colloquial phrases, usually in one sentence, stating a role, an objective or wish and ideally the aspired benefit. Example:

As a traveler I want be able to check the availability of different flight connections to my destination for finding the best flight for me.

The set of stories still to be implemented is called product backlog. Those planned for the current sprint are called the sprint backlog. Stories should only be assigned to a sprint backlog at the beginning of a sprint. Due to short sprint durations, this allows a maximum flexibility for the selection of requirements to be met by the next product increment. Moreover, the product backlog can be extended, reduced or stories can be re-prioritized without impact on the current development process.

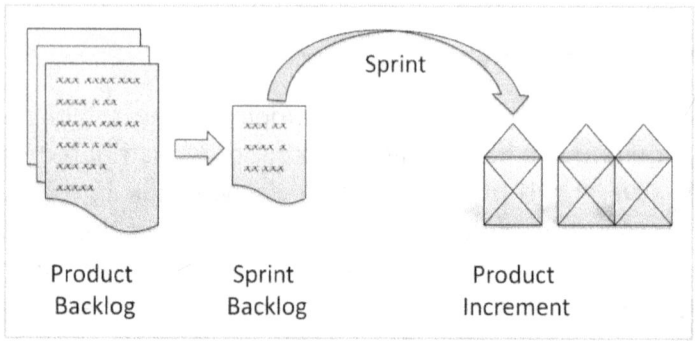

Figure 1: Principle of Incremental Product Development

In case of a fixed delivery deadline and contractually agreed requirements, team size remains the most important factor for planning. Thus, the manpower required for implementing the stories or requirements is estimated first and then, depending on the available time, the required team size can be determined. An exemplary estimation that results in a required manpower of 500 man days (md) with 100 working days remaining means that a team of at least five persons is required – plus a safety margin.

Expert Estimations

A proven method to approximate the costs for implementing a user story is the expert estimation. One or more experts independently estimate based on previous experience with similar tasks. Since individual experience can differ to a high degree and depends on the own capabilities, it is obvious to use the average over all estimations. In addition to this subjectivity, another problem of the method is that experts mostly are members of development teams and cannot productively contribute to the development process while performing their estimation.

A derivative of the expert estimation is the Delphi method, where a moderator first presents details of the user stories to be estimated to all involved experts. Subsequently, they perform an individual estimation, completely independent from each other. Then the moderator evaluates, presents the results and thereby points out discrepancies – all with respect to anonymity. Discussions are undesirable to prevent an influence on group dynamics through dominant experts. Every expert has the opportunity to rethink and change his estimation. This process is repeated until the discrepancies do no longer exceed a tolerance threshold. An advantage of the Delphi method is the iterative refinement of all expert estimations and the resulting higher accuracy. A disadvantage is the high time commitment of the experts.

Indirect Estimations with Story Points

Any direct estimation depends on the estimator's individual skills and experiences and does usually not consider the capability of the team planned for the development. Contrary to this, indirect methods first determine the size of the objects respectively requirements to be developed and then put it in relation to empirical values of, ideally, the same team from earlier sprints.

In case of the story points method, estimators rate the size of user stories relative to each other and not in an absolute unit of measurement. Example: "The selection of the destination airport and the date is a one. Relative to this, the query for available flights is a three. Accordingly, booking a selected flight is an eight." The scale of story points is aligned to the Fibonacci sequence: 1, 2, 3, 5, 8, 13, and so on. [Cohn 2013]. This avoids many discussions about details and makes decisions between the valid numbers easier. At the same time it is obvious that, using this scale, the accuracy of estimations decreases with the increasing size of stories.

The calculation of the prospective development costs on basis of story points also requires an empirical value of how many story points the team can develop per increment or sprint under comparable conditions. This empirical value is called velocity and is updated after each completed sprint by determining the size of all successful developed stories in relationship to the required time.

Advantages of the story point method are its quick use and that it is being based on the current team performance instead of the ability of individual experts. Due to the periodical recalibration, velocity can be fit to changes in team composition or other basic conditions within few sprints.

A disadvantage is the detail level of the object to be estimated: a user story which describes a requirement by only one sentence. Any unexpected complexity first discovered during the implementation often results in exceeding the estimated effort.

Indirect Estimations by measuring the Functional Size

A different type of an indirect estimation is determining the functional size (instead of story points) and relating it to an empirical value describing which size can be implemented with specific personnel costs (often called productivity or efficiency). The basics for measuring the functional size are defined in the industry standard ISO/IEC 14143 [ISO/IEC 14143 2007]. They require the refinement of user stories to use cases.

A use case stands for a system behavior perceptible by an actor from outside of the system boundaries. Actors can be human system users as well as other systems or machines (hardware or software). An example is the use case diagram of a (highly simplified) Internet Booking Engine for flights shown in figure 2. Within the system boundaries (shown as a rectangle with the label "Internet Booking Engine") use cases are represented by ellipses. They are related to (used by) actors, which in the example below are a traveler and a Computer Reservation System (CRS). Their relation is indicated by lines.

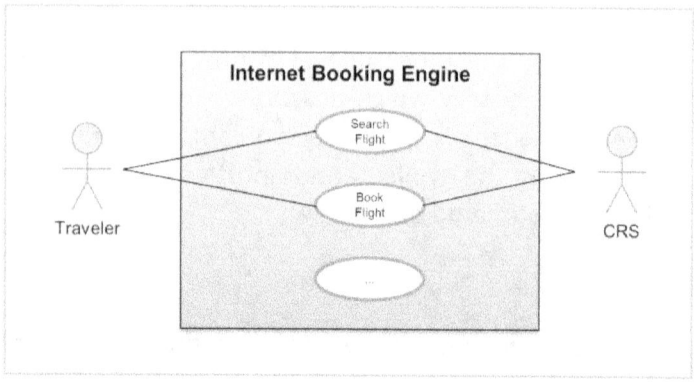

Figure 2: Sample Use Case Diagram

Each use case stands for actions which can be described by base functional components (BFCs) respectively elementary processes. The previously shown exemplary user story

> *As a traveler I want be able to check the availability of different flight connections to my destination for finding the best flight for me*

could, as a use case "search flight", consist of the following (simplified) elementary processes:

1.	Traveler calls the dialog Flight Search.
2.	Traveler enters the departure date.
3.	Traveler enters the first letters of the destination (name or code).

4.	System looks up for matching airports in the database and displays a list of found items showing names and codes.
5.	Traveler selects a list entry and clicks on the button "Search".
6.	System sends a message of type "Flight Search Request" including the departure date and airport code to the CRS.
7.	System receives a message of type "Flight Search Response" from the CRS and reads the fields departure time, arrival time, airline, flight number, class, price and currency of all flight records included.
8.	System displays a table showing this information about all received flight records.

The description of these details is a mandatory requirement for the determination of the functional size. Since it is also a mandatory requirement for the implementation, this work means no additional effort.

Summary

Wherever an initial and quickly available rough planning is required, methods such as story points, which can be applied rapidly, are appropriate. After use cases have been described, this rough planning then can be refined with more detailed information by applying methods which will be described in the next chapter.

Methods conformal to the standard ISO/IEC 14143 allow the determination of functional size according to precisely defined rules without any room for interpretation. In the following chapter, applying the Function Point Analysis, the COSMIC method and the Data Interaction Point method will be explained on basis of the exemplary use case already shown before. The subsequent chapter describes how the empirical values required for an indirect estimation can be measured and which information can be derived from its chronological course.

Methods for Functional Size Measurement

Function Point Analysis

The best-known method for measuring the functional size, the Function Point Analysis (FPA), has been developed by Allan J. Albrecht at the end of the 70s. It counts the elementary processes of use cases and the related data objects. Regarding elementary processes, the following types must be differentiated:

- **External Input** (EI). In the contest of a use case, data exceeds the system boundaries from the outside.

- **External Output** (EO). In the context of a use case, data exceeds the system boundaries to the outside.

- **External Inquiry** (EQ). Data, for example search criteria, exceeds the system boundary from outside and triggers a query process, whose results exceed the system boundary to the outside.

When counting, elementary processes are weighted according to the number of involved data element types (DETs) and the number of file types referenced (FTRs). These weightings are defined by matrices with three-tiered open-ended interval scales (see figure 3).

External Input

Data Element Types (DETs)

EI	1-4	5-15	≥16
0-1	3	3	4
2	3	4	6
≥3	4	6	6

File Types Referenced (FTRs)

External Output

Data Element Types (DETs)

EO	1-5	6-19	≥20
0-1	4	4	5
2-3	4	5	7
≥4	5	7	7

File Types Referenced (FTRs)

External Inquiry

Data Element Types (DETs)

EQ	1-5	6-19	≥20
1	3	3	4
2-3	3	4	6
≥4	4	6	6

File Types Referenced (FTRs)

Figure 3: Weighting Matrices (1) of the Function Point Analysis

When counting data objects which are involved in the elementary processes, the following types must be differentiated:

- **Internal Logical File** (ILF). This is usually a data structure stored in a database which is maintained by the system examined. Hence, insertions, deletions and updates belong to the use cases of this system.

- **External Interface File** (EIF). This is a data structure stored in a database to which the considered system only has read access. Therefore, insertions, deletions and updates are none of its use cases.

The weight of a data object depends on its number of data element types (DETs) and its number of record element types (RETs). A record element is a group of related data elements. Examples are the name of a person, which consists of the data elements salutation, title, first name and surname, or the address, which in the simplest case consists of the data elements zip code, city and street. Thus, a personal postal address can be an ILF build by the two record element types Name and Address with a total of 7 data element types. Also for data objects the weights are defined by matrices with three-tier open-ended interval scales (see figure 4). Thus, the weight of the sample ILF with 2 RETs and 7 DETs is 7 function points (FP).

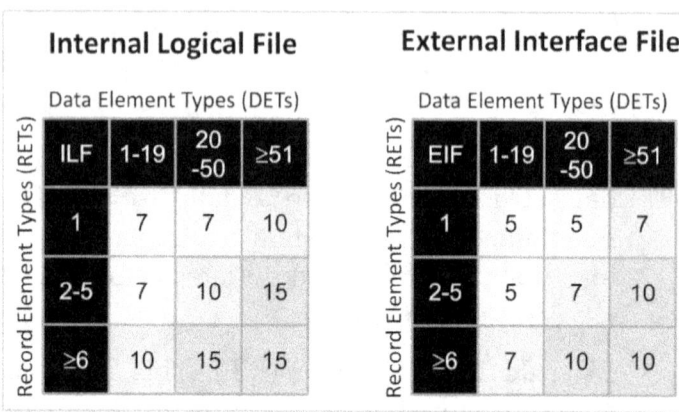

Figure 4: Weighting Matrices (2) of the Function Point Analysis

The previously described method has been standardized by the International Function Point Users Group (IFPUG) in the industry standard ISO/IEC 20926 [ISO/IEC 20926 2009]. There are different variations of this method which are also documented in ISO standards, for example the Mark II FPA method of the UKSMA (United Kingdom Software Metrics Association) or the methods of FISMA (Finnish Software Measurement Association) and NESMA (Netherlands Software Metrics Users Association). However the IFPUG method is the world's best-known variant and is regarded as industry standard for functional size measurements.

For the sample use case "search flight" described in the previous chapter, the determination of function points is as follows:

1. Traveler enters the departure date and the first letters of the destination (name or code): 1 external input (EI) with 1 FTR and 2 DETs = 3 FP

2. System looks up for matching airports in the database and displays a list of found items showing names and codes:
   1 external interface file (EIF) with 1 RET and 2 DETs = 5 FP

3. Traveler selects a list entry and clicks on the button "Search":
   1 external input (EI) with 1 FTR and 1 DET (the ID of the selected list entry) = 3 FP

4. System sends a message of type "Flight Search Request" including the departure date and airport code to the CRS:
   1 external output (EO) with 1 FTR and 2 DETs = 4 FP

5. System receives a message of type "Flight Search Response" from the CRS and reads the fields departure time, arrival time, airline, flight number, class, price and currency of all flight records included:
   1 external input (EI) with 1 FTR and 7 DETs = 3 FP

6. System displays a table showing this information for all received flight records:
   1 external output (EO) with 1 FTR and 7 DETs = 4 FP

The total result of this calculation is a functional size of 22 function points (FP).

If an empirical value of the productivity is available from last retrospective, which is the information how many function points can be developed on average per work unit, then the expected effort in terms of personnel costs can be calculated:

$$\text{Effort} = \frac{\text{Functional Size}}{\text{Productivity}}$$

In case the average productivity is 5.5 FP/MD (function points per man day), the implementation of the use case "Search Flight" with a functional size of 22 FP results in an expected effort of 4 man days – for the same activities and under the same conditions as this productivity value has been measured.

COSMIC Method

The idea to count data elements directly instead of counting elementary processes and using the number of involved data elements only for weighting lead to the development of the Full Function Points method (FFP method) in the 80s. Based on this, the Common Software Measurement International Consortium (COSMIC) has been founded in 1998 [COSMIC 2015]. In 2003 the FFP respectively COSMIC method was accepted as industry standard ISO/IEC 19761 [COSMIC FSM 2014].

The COSMIC method counts data movements which are data elements involved in the elementary processes that exceed the system boundaries respectively are stored in or read from the database. It distinguishes the following types of data movements:

- **Entry**: One or more data elements coming from an actor exceed the system boundaries and are used by a functional process of the system. Instead of actors, COSMIC also uses the expression of functional users, which, in addition to human users or external systems, can denote any hardware or software acting as an input or output device.

- **Exit**: One or more data elements of a functional process exceed the system boundaries towards an actor respectively functional user.

- **Read**: One or more data elements are read from persistent memory and used by a functional process.

- **Write**: One or more data elements from a functional process are stored in persistent memory.

The functional size is derived from the number of data movements categorized as entry, exit, read or write – without weighting. For the previously used sample use case "search flight", the determination of the functional size according to the COSMIC method is as follows:

1. Traveler enters the departure date:
   1 x Entry

2. Traveler enters the first letters of the destination (name or code):
   1 x Entry

3. System looks up for matching airports in the database and displays a list of found items showing names and codes:
   2 x Read

4. Traveler selects a list entry and clicks on the button "Search":
   1 x Entry

5. System sends a message of type "Flight Search Request" including the departure date and airport code to the CRS:
   2 x Exit

6. System receives a message of type "Flight Search Response" from the CRS and reads the fields departure time, arrival time, airline, flight number, class, price and currency of all flight records included:
   7 x Entry

7. System displays a table showing this information about all received flight records:
   7 x Exit

In total these are 21 data movements. Therefore the functional size is 21 COSMIC function points (CFP).

Also in this case, the implementation effort is the quotient of the functional size determined as shown before and an empirical value of the average productivity. Example: In case that the average productivity is 5 CFP/MD (COSMIC function points per man day), the resulting effort is 4.2 man days - provided that the general conditions are comparable between the development project(s) of which the empirical value was measured).

Data Interaction Point Method

In 2006, PASS Consulting Group was in search of a functional size measurement method where data elements and not elementary processes would be counted, but - in contrast to the COSMIC method - the weighting would scale with the complexity of the functional processes. As a result from this, the Data Interaction Point method (DIP method) was developed [PASS 2014]. This method distinguishes the following categories of data elements:

- **User interface, dialogs/masks** (DIP-UI): In case of modern, dialog-oriented web systems, input elements crossing the system boundaries coming from an actor or functional user are counted with a value of 3, while output elements crossing the boundaries towards an actor or functional user are counted with a value of 1. According to the technology, these weights can be adjusted, for example in case of character-oriented input masks with less complex input functionality.

- **Interfaces, import or export functions** (DIP-IMP/EXP): This category aggregates all data elements which cross the boundaries, for example as import or export, or via an interface to an external system, and are used by a functional process of a use case. Each element is counted with a value of 1.

- **Database** (DIP-DB/REF): Data elements which are only read by the regarded system are counted with a value of 1. Those with write access by the system are counted with a value of 3.

For the previously used sample use case "search flight", the functional size according to the Data Interaction Point method is determined as follows:

1. Traveler enters the departure date:
   1 UI input element x 3 = 3 DIP
2. Traveler enters the first letters of the destination (name or code):
   1 UI input element x 3 = 3 DIP
3. System looks up for matching airports in the database and displays a list of found items showing names and codes:
   2 elements in a reference table (DIP-REF) x 1 = 2 DIP
4. Traveler selects a list entry and clicks on the button "Search":
   1 UI input element (ID of the selected list entry) x 3 = 3 DIP
5. System sends a message of type "Flight Search Request" including the departure date and airport code to the CRS:
   2 export/ interface elements (DIP-EXP) x 1 = 2 DIP

6. System receives a message of type "Flight Search Response" from the CRS and reads the fields departure time, arrival time, airline, flight number, class, price and currency of all flight records included:
   7 import/ interface elements (DIP-IMP) x 1 = 7 DIP

7. System displays a table showing this information about all received flight records:
   7 UI-output elements (table columns) x 1 = 7 DIP

This results in a functional size of 27 data interaction points (DIP).

Assumed that the empirical value for the average productivity of 6 DIP/MD (data interaction points per man day), the effort resulting from this sample is 4.5 man days – comparable development conditions provided.

Extending Methods for measuring the Size of Further Development

All previously described examples for the determination of the functional size concern new developments. In case of further developments, an indirect cost estimation must not only consider new functional processes to be implemented but also existing ones which have to be changed or deleted. Depending on the measurement method, the question is which values must be assigned to changed or deleted objects. Regarding the function point analysis and the data interaction point method it is proven practice to count them with the lowest value in the according category. For FPA, this is 3 for external inputs, 4 for external outputs, and etcetera, whereas the DIP method generally attributes a value of one.

The Impact of Complexity

In the context of software development, three types of complexity can be distinguished, all having different impacts on the reliability of cost estimations.

The Complexity of an Implementation

This type of complexity has an impact on the effort to understand the code of an existing application to be developed further. Reasons for a high complexity can be, for example, bad coding style, insufficient commenting or weaknesses in the application design. The consequence is a higher expenditure of time required for implementing of new requirements.

For the size metrics described before this type of complexity is not an issue, as they consider functional requirements and are independent from technical aspects. To estimate the costs of further developments, it is a good practice to use such an empirical productivity value that has been measured at previous development cycles of the same application. This allows the specific complexity of the application to directly flow into the estimation.

Interactional Complexity

All methods introduced in this chapter consider interactions of the related system with the actors of its use cases (its functional users). Therefore it is obvious that there is an influence by the complexity of these interactions respectively the functional processes which are triggering the interactions or are triggered by them.

The function point analysis considers this interactional complexity by placing different weights on the counted elementary processes and the counted data objects which cross the system boundaries. As explained before, the weight is derived from the number of related data elements respectively structures, which undoubtedly correlates with the complexity.

A different approach is to derive the complexity from the type and direction of data movements, as implemented by the data interaction point method. The input of a data element into a dialog usually has a higher complexity than the output, due to the required validation of entered values and the persistence. Therefore it is counted with a higher value. This is analogous to writing a data element into the database, which needs to be validated, checked for consistency and integrity and is therefore more complex than read access. The advantage of these distinctions is that usually the types of data elements to be counted can be distinguished and assigned easily.

Algorithmic Complexity

Wherever the main focus is on interactions of a system with human users or external systems, it is reasonable to measure the functional size on a use case basis, as it has been described in the standard ISO/IEC 14143. This measurement considers only the data movements between the system to be measured and the actors of the use cases, together with the representation of the related data in the database.

If the main focus is not on interactions but on the execution of complex algorithms, the results of these measurement methods only have limited reliability. This may become clear in case of a route planning software, which only receives a start and a destination address as input data and provides a list of route sections as output data. Measuring the functional size might result in a very low value. As a consequence, an indirect cost estimation would calculate also a low development effort. Too low, because route planning algorithms are complex and the effort for implementing them is rather high.

Therefore indirect cost estimations based on the functional size are not recommendable if the focus is on complex algorithms or business logic. In such cases, expert estimations might achieve better results.

Method Comparison

The effort for functional size measurements depends strongly on the structure of the functional requirements, the quality of their descriptions and on own methodological expertise. The samples introduced in the previous chapters have shown that, due to the application of interval scales, the function point analysis (FPA) requires more steps than other methods directly counting data movements or data elements. In practice, this sometimes is compensated by approximation methods which ignore the interval scales and use constant approximation values instead. This reduces the measuring effort at the price of entirely neglecting complexity.

The fact that the common version of the FPA, as standardized by the IFPUG, takes interaction complexity into consideration, is an advantage of this method. This is also the case with the data interaction point method, however not with the COSMIC method, which attributes the same value to each data movement. None of these methods consider algorithmic complexity.

Looking at the measuring accuracy, FPA has the issue that the weights of counted elementary processes and data structures do not increase over a certain size. For example, internal logical files (ILFs) with multiple record element types (RETs) always are counted with the value of 15 if the amount of data element types (DETs) is higher than 50 - notwithstanding the fact that data structures of old systems often have several hundred different fields. The COSMIC and the data interaction point method have no such limitation: Every data movement respectively every data element increases the functional size.

Further Methods

In addition to the three methods introduced before, there are numerous other ones which more or less consider functional requirements. Some of them are listed below – in alphabetical order and without any claim to completeness:

- Bang Metrics, De Marco, 1982. Based on the structured analysis, so-called functional primitives are counted. Weights are derived from the numbers of input and output tokens.

- Data Point Method, Sneed, 1989. Counts tables, keys, relations and attributes in the database and their representations in dialogs and interfaces. Weights are estimated.

- FISMA Function Point Method (ISO/IEC 29881), Finnish Software Measurement Association (FiSMA), 2009. Counts data elements crossing the system boundaries, their use in algorithms and read/write access to the database. Weights are use-related.

- Mark II FPA Method (ISO/IEC 20968), United Kingdom Software Metrics Association

(UKSMA), 1998. Based on FPA. Counts elementary functions and access to the database using defined weight values.

- NESMA Function Point Method (ISO/IEC 24570), The Netherlands Software Metrics Users Association (NESMA), 2005. Additionally to approximation methods, it includes the method "Detailed FPC", which is identical to the original FPA.

- Object Point Method, Sneed, 1994. Based on a class model, it counts classes, processes and messages.

- Use Case Point Method, Karner, 1993. Counts use cases and actors. Derives the weights from three-tier interval scales.

If costs are estimated indirectly, the functional size of the requirements to be implemented must be determined with the same method used to measure the related empirical value of the own productivity. Therefore, a company must opt for one method, taking into consideration the effort of applying it, existing expertise, if required also the ability to calibrate weight values for different system types, the suitability for different system sizes and the management of complexity.

Measuring the Reference Value for an indirect Cost Estimation

The previous chapter has shown how the size of functional requirements can be determined. In case of indirect cost estimations this value is divided by an empirical value of the development productivity measured under comparable conditions. The following chapter addresses the measurement and steady update of this reference value, related requirements and possible interpretations.

Regular Measuring of Productivity

The principle of indirect cost estimating requires an empirical value of own productivity. It can be calculated after a completed development process, for example within a retrospective, using the following formula:

$$\text{Productivity} = \frac{\text{Functional Size}}{\text{Effort}}$$

"Functional Size" is the size of the software implemented and tested successfully in the course of the development process. "Effort" is the total personnel costs of this process.

Process Scope regarding Work

Begin and end of the considered development process must be clearly defined. Only those personnel costs must flow into the calculation which are related to a task of the process – not automatically any work done by members of the project team. Those activities not related to the development process must not be considered – this also shall apply for analyzing and fixing of defects from previous releases. Then again, work done by non-team-members for supply of the development process, in most cases also for supporting it, must be considered. If this rule is not adhered to, the calculated productivity is not suitable for reliably estimating the development costs for new increments.

Process Scope regarding Sub-Processes

The results of indirect cost estimations based on an empirical value of productivity are always related to the same sub-processes which have been considered when measuring this empirical value. For example, if the measurement of the reference value includes work beginning with the end of the description of use cases, any indirect cost estimation derived from it also results in the expected effort for the development process after use cases are described. As a principle, the process scope of a productivity measurement can be tailored as needed. Therefore it is possible to explicitly measure the productivity of the conceptualization or the implementation phase, of a specific test stage or of any other sub-process - provided that personnel costs can be clearly assigned to the related activities, which in practice can be more of an organizational challenge. If such particular measures of productivity would be available, an organization could – after determining the functional size to be developed – calculate the personnel costs independently for each sub-process respectively work type.

Process Scope regarding Quality

As important as defining the start of the process, which is in scope for productivity measurements, is its end. This can be determined by end criteria of the observed sub-processes regarding the achieved quality of its artefacts - a "quality gate" with precisely defined rules and criteria. If required, the productivity measurement may have to be restricted on these software components which meet the demanded criteria and will be used for the next product increment.

Automated Measurements

Mapping of Objects to be counted on Structural Characteristics

The post-implementation measurement of functional size can be time-consuming. Moreover, this work is not very creative - often regarded as bean counting - and its benefit after the successful development of a new product increment often seems to be highly questionable. Nevertheless, this work is crucial for the reliability of future cost estimations.

All three methods introduced in this book count objects whose equivalents can be found in the source code, in metadata or in models of the according software. For identifying these objects, mostly clear rules can be found depending on design and naming conventions. In these cases a program can be implemented which parses the sources and counts objects which meet the rules. It even can keep an inventory of these objects. Some examples for such automation approaches are:

- GUI frameworks which often keep models or templates of the dialogs, technically mostly in XML or XHTML files. If tags can be assigned unambiguously to specific types of dialog elements, these files can be automatically parsed by a program and analyzed regarding the frequency of these tags.

- Interfaces based on XML schemes or WSDL descriptions (Web Services Description Language). Here as well, particular tags may be assigned to the data elements which then must be counted allowing for an automatic analysis by a program. Regarding this, it is important to consider that not all data elements defined within the scheme of a standard interface may be used by the considered use cases. Only these data elements actually used have to be counted, not the entirety of those included in the data structure.

- Tables, attributes and relations in a database are stored as metadata, can be read from system tables of the DBMS (Database Management System) and analyzed by a counting program.

Of course, a counting program tailored to a specific system does require a not insignificant effort for the analysis and its implementation. But once created, it enables the team to measure the functional size by a simple program call. By comparing the result with the value measured for the previous release, the size of newly implemented functionality can be determined. Then the outcome can be used along with the personnel costs for calculating productivity.

Possible Restrictions

One problem of the automated measurement of functional size is posed by use cases not always being mapped one-to-one to structural characteristics of the software. Good programming style and good design often implicate that objects are implemented once and then re-used multiple times. A dialog which is used by several use cases may serve as an example. Suppose that in a system A this dialog is implemented once, in a way that its functionality can be controlled by the business logic, directly enabling its use by all according use cases. In another system called system B, this dialog is implemented once per use case, or implemented once and then copied and modified as required. A functional size measurement considering the implemented dialogs would measure a smaller size for system A, in which the dialog is implemented once, than for system B, in which the dialog exists multiple times with all implementations being almost identical to each other. Actually, from a methodological point of view, the higher value measured for system B is the correct one for both systems, because the measurement method, if conformal to the standard ISO/IEC 14143, considers use cases which are represented in system B by the different copies of the dialog. The lower value measured for system A, characterized by quite a few use cases using the same dialog, is too low because the measurement did only consider one use case.

Unfortunately, it is difficult in practice to include the aspect of reuse into automated functional size measurements. As long as measurements of the same system are used to calculate only the growth compared to the previous increment, this issue can mostly be neglected. However, comparability of size values between different systems which were measured based on structural characteristics instead of use cases, is very limited in case these systems have different levels of reusability.

Iterative Refinement of the measured Productivity

In practice, it can often be observed that, in case of smaller increments respectively releases, the measured productivity deviates more from the mean than in case of larger ones. Mostly, the reason is that the complexity of implemented requirements can be highly variable and is not completely taken into consideration by the measurement methods described before, which can result in partial inaccuracies when measuring smaller sizes.

Two examples will explain these circumstances: Release A only includes some new implemented reports which represent a large number of data elements crossing the system boundary. These data elements result in a significant growth of the measured functional size. If the report data can be read from the database by simple queries, the development effort might be small. A large increase in functional size implemented with little effort results in a high productivity for release A.

Let us assume that in release B, a complex algorithm, which is to output its result in only one dialog field, has been implemented. All methods described before measure a small functional size, because only one data element crosses the system boundary and can be noticed by an actor. Due to the algorithmic complexity, the development effort may be high. A small growth of the functional size implemented with great effort results in a low productivity for Release B.

When measuring small releases or increments, there is a risk that the few implemented requirements deviate highly from the mean complexity and therefore the calculated productivity becomes partially inaccurate. In case of greater releases, these discrepancies are usually leveling out. Therefore it is a good practice for measuring the average productivity of the further development of a specific system, if it is required as a reference value for indirect cost estimations of new releases, to summarize the considered functional size and effort over an extended period:

$$\text{Average Productivity} = \frac{\sum \text{Functional Size}}{\sum \text{Effort}}$$

Example: In table 1, the implemented functional size S_n (measured in data interaction points) and the effort E_n (in man days) are listed for several releases of the same system.

	Functional Size S_n (DIP)	Effort E_n (MD)	Average Productivity $\bar{P}$ (DIP/MD)
Rel. 1.0	125	23	
Rel. 1.1	23	2	
Rel. 1.2	540	123	
Rel. 1.3	125	15	5.0
Rel. 1.4	410	149	3.8
Rel. 1.5	90	77	3.2
Rel. 2.0	1.210	85	5.6
Rel. 2.1	435	26	6.4
Rel. 2.2	100	88	6.6
Rel. 2.3	995	200	6.9

Table 1: Sample Performance Indicators of Further Development (1)

In this example, each average productivity value (listed in the last column) is calculated based on the sum of functional size and the effort of the last four releases. Thus, the average productivity of release 1.3 is calculated as follows:

$$\bar{P} = \frac{125 + 23 + 540 + 125}{23 + 2 + 123 + 15} = 5$$

This way of calculating productivity is resilient towards outliers as release 1.1, in which an individual calculation shows a productivity of 11.5 (resulting from a size of 23 DIP divided by an effort of 2 MD), or as release 2.2 with an individually calculated productivity of 1.1 (100 DIP divided by 88 MD). Only when calculating productivity by summarizing multiple measurements, sustainability of the productivity improvement, which in this example is effective since release 2.0, becomes apparent. This is not transparent when looking at single measurements only.

By taking the sum over 4 releases, as done in this example, the described leveling effect can be explained sufficiently. Depending on the release cycle, it can be beneficial to use even longer periods, for example a complete year.

Considering non-functional Requirements

When developing new IT systems, it often is difficult to estimate the costs for implementing non-functional requirements. Below are some examples, for which the costs may not be insignificant:

- a large number of concurrent users,
- a high transaction volume to be processed by the considered system or to be exchanged with other systems,
- the continuous availability of the system without noticeable downtimes even in case of serious failures,
- barrier-free use of all system functions.

In many cases, non-functional requirements can be implemented by using the right architecture or scaling and configuring it accordingly. Examples for this are performance requirements, which in an ideal case can be met by a sufficient number of load balanced servers. While personnel costs are low, there are infrastructural costs and efforts for performance testing. As follows, the implementation of non-functional requirements cannot be measured directly with methods as they have been described before. It requires architectural expertise and should therefore be estimated separately, for example in the course of an expert estimation.

However, the impact of non-functional requirements on the development costs for functional requirements can be considered easily. It is only required that the indirect cost estimation uses a reference value of the productivity measured under comparable conditions. In case of the further development of an already implemented system, this is not an issue if the reference value has been measured based on the previous releases and therefore it considers the inherent system properties.

Regular Measurements

The empirical value of the own productivity used as reference for indirect cost estimations must be ascertained cyclically and hence updated regularly. A potential program for the automated measurement of functional size would calculate the difference between the current and the previous size measurement, which must be put into relation to the relevant personnel costs. In the ideal case this is a task of a few minutes and an integral part of each retrospective.

The opportunity of reliable cost estimations for new releases is only one advantage of always having the latest productivity value at hand. Another one is the ability to quickly calculate the impact of planned or unplanned changes of the development conditions. Furthermore, there is the benefit of a transparent time course of productivity and quality as well.

The Relationship of Productivity and Quality

When measuring the development productivity of a release or increment as it is required for indirect cost estimations, the effort for analyzing and fixing defects from previous releases does not flow into the calculation. It can be concluded that, solely based on such productivity values, it is not possible to rate improvements or deteriorations in quality. While considering defect costs additionally to the development costs would reduce the productivity value in case of quality deficiencies, these values are not suitable for indirect cost estimations of new requirements because defect costs do not arise continuously.

It is proven to be more advantageous to measure productivity and quality separately and then compare both figures with each other. Thereby a simple quality indicator can assume the number of 'real" defects, meaning defects in terms of unfulfilled requirements or undesirable system behavior – no new requirements disguised as defects or even handling errors are considered. Alternatively, a quality indicator can be determined by the effort for analyzing and fixing defects – similar to the development productivity.

	Productivity $\bar{P}$ (DIP/MD)	Quality Q (Number of Defects)
Rel. 1.3	5.0	10
Rel. 1.4	3.8	6
Rel. 1.5	3.2	4
Rel. 2.0	5.6	6
Rel. 2.1	6.4	10
Rel. 2.2	6.6	6
Rel. 2.3	6.9	4

Table 2: Sample Performance Indicators of Further Development (2)

Table 2 shows the sample productivity values of the different releases discussed in the previous chapter, supplemented by the assumed number of production defects per month, which serve as quality indicators. After inserting these figures into a XY diagram, the course of the graph mostly shows long-termed trends (see figure 5). In this example, the productivity first decreases until after release 1.5 there is a turning point leading to a sustained improvement of the productivity. At the same time the large release 2.0 also includes new defects which temporary reduce quality. From release 2.1 onwards, the number of defects decreases and a trend into the right direction is apparent: productivity and quality as well are improving continuously.

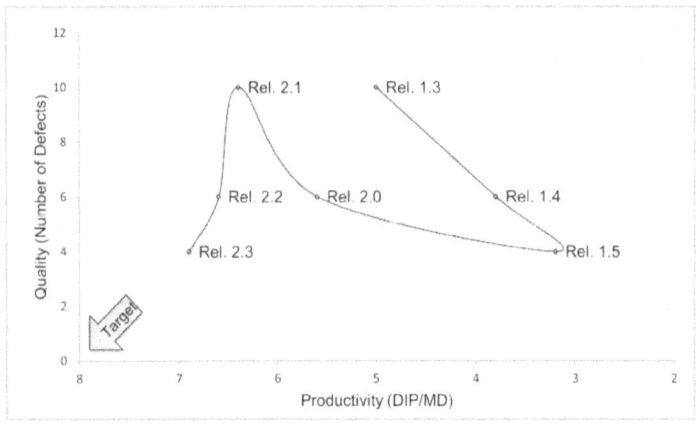

Figure 5: Course of Productivity and Quality in Further Development

A comparison of productivity and quality can reveal neglects of quality assurance measures, for example if, due to reduced testing effort, the graph points to an increase in productivity with continuously deteriorating quality. The trend of the graph observed over several measuring points can also show how effective improvement measures have been, especially in which period of time, and how sustainable the improvement of productivity and quality has been.

Conclusion

Agile software development and a binding agreed delivery date are not mutually exclusive at all. On the contrary, the principles of the Agile Manifesto are well-proven measures to mitigate the typical risks of development projects with binding regulations.

Crucial for the achievement of project goals is a reliable plan which can be created without significant expense and readjusted flexibly in case of changed requirements. Under these conditions, indirect cost estimations on basis of functional size measurements are a proven practice. They require the size of the functional requirements as an input value, determined by precise rules, as well as an empirical value of the own productivity, measured after earlier sprints.

For the determination of functional size, methodological knowledge and expertise is required regarding the selected method, for example Function Point, COSMIC or Data Interaction Point method, along with a sufficient specification of requirements, for example by described use cases and elementary processes. Under these conditions, the resulting value is a valid key figure of the functional requirements' size. It is independent from the individual performing the estimation. Repeated estimations of the same requirements always result in the same value.

An empirical value of the own productivity, that is the functional size which can be typically implemented by the team with given personnel costs, is required in addition. It can be measured and updated within the scope of regular retrospectives of completed increments, sprints or releases. Thereby, the implemented functional size is set in relation to the arising personnel costs. The functional size is determined by counting – depending on the method – elementary processes or data elements. With an appropriate mapping to structural characteristics of the system, this counting process can be automated. A such counting program, once implemented, can quickly measure the total size of the system and, by comparing the current size with the previous measurement, return the growth of the functional size resulting from the last development process.

Regularly measured and compared values of productivity and quality - in the simplest case the number of production defects – aid in recognizing any need for action and allow to verify the effectivity of implemented improvement measures.

Glossary

- **Actor**: User or external system interacting with a target system in the context of a → **use case**.

- **Algorithm**: Finite sequence of executable instructions to solve a problem.

- **Algorithmic Complexity**: The → **complexity** of the program logic implemented in an application.

- **Complexity** [in software development]: The effort required to understand a program or → **algorithm**.

- **Crystal**: A class of methods for agile software development.

- **Elementary Process**: Also: Base Functional Component (BFC). A single step in a → **use case**.

- **Extreme Programming**: Methodology for → **iterative** software **development**.

- **Fibonacci Sequence**: Infinite sequence of natural numbers starting with number 1 for two times and continued by each the sum of two consecutive numbers: 1, 1, 2, 3, 5, 8, 13 …

- **IFPUG**: Abbreviation for the International Function Point Users Group, a non-profit worldwide organization for the standardization and promotion of the function point analysis.

- **Incremental Development**: Type of software development where parts of a system are developed at different times and the system is extended by already completed parts.

- **Interactional Complexity**: Complexity of → **use case** related interactions by → **actors** with a considered system.

- **ISO/IEC 25010**: Standard for software quality. Defines software quality characteristics and their partition into sub-characteristics. Formerly: ISO/IEC 9126.

- **Iterative Development**: Step-by-step refinement of implementing → **requirements**, often starting with sophisticated and risky requirements and approximating the system closer to the objective with each iteration.

- **Non-functional Requirement**: Also: Non-functional User Requirement, NFUR. Specifies the expected characteristics of a product according to → **ISO/IEC 25010**.

- **Product Backlog**: List of → **requirements** to be implemented and defects to be fixed – prioritized by the benefit from the product owner's point of view.

- **Productivity**: A process metric. In software development, productivity is usually calculated as the ratio of the developed software's size and the effort for its development.

- **Quality Gate**: A milestone in the course of a project where continuation or completion depends on the compliance with defined quality criteria.

- **Requirement** [in software development]: Agreed or expected characteristics of a system – separated into functional and → **non-functional requirements**.

- **Risk** [in development projects]: A threat to objectives which can be rated by its probability of occurrence and the potential extent of loss.

- **Scrum**: Methodology for → **iterative** and → **incremental software development** in short cycles (→ **sprints**) which follows agile principles and knows three roles: a product owner who is responsible for the product vision, team members and a scrum master who is a moderator experienced with the methodology.

- **Sprint**: Cycle of an agile software development process with a typical duration between two and four weeks.

- **Sprint Backlog**: Selection of entries from the → **product backlog** which are up for implementation during a → **sprint**.

- **Story, User Story**: Colloquial phrase, usually in one sentence, stating a → **requirement**.

- **Use Case**: Use cases describe all scenarios how → **actors** can accomplish specific goals by using a considered system. Each use case is defined by → **elementary processes** defining interactions between actors and the system abstracted from specific technical solutions.

- **Virtualization**: Simulation of a physical object or a resource using IT.

Bibliography

- **[Agile Manifesto 2001]**: Website "Manifesto for Agile Software Development". URL http://agilemanifesto.org/iso/en/ (29.12.2015).

- **[Agile Manifesto Principles 2001]**: Website „Principles behind the Agile Manifesto ". URL http://agilemanifesto.org/iso/en/ principles.html (29.12.2015).

- **[Cohn 2013]**: Website "How Can We Get the Best Estimates of Story Size?". URL https://www.mountaingoatsoftware.com/ blog/how-can-we-get-the-best-estimates-of-story-size (01.11.2015).

- **[COSMIC 2015]**: Website of the Common Software Measurement International Consortium. URL http://www.cosmicon.com (11.02.2015).

- **[COSMIC FSM 2014]**: "The COSMIC Functional Size Measurement Method Version 4.0; Measurement Manual; The COSMIC Implementation Guide for ISO/IEC 19761:2011".URL http://www.cosmicon.com/portal/public/MM4.pdf (11.02.12015).

- **[Forrester 2005]**: C. Schwaber, R. Fichera (2005): "Corporate IT Leads The Second Wave Of Agile Adoption". Forrester Research Inc.

- **[ISO/IEC 14143 2007]**: "Information technology -- Software measurement -- Functional size measurement -- Part 1: Definition of concepts". ISO (International Organization for Standardization).

- **[ISO/IEC 20926 2009]**: "Software and systems engineering -- Software measurement -- IFPUG functional size measurement method 2009". ISO (International Organization for Standardization).

- **[PASS 2014]**: S. Luckhaus (2014): "Produktivität in der Softwareentwicklung: Band 1 - Produktivitäts- und Leistungsmessung - Messbarkeit und Messmethoden". PASS IT-Consulting Dipl.-Inf. G. Rienecker GmbH & Co. KG.

- **[VersionOne 2013]**: "7th Annual State of Agile Development Survey". URL https://www.versionone.com/pdf/7th-Annual-State-of-Agile-Development-Survey.pdf (01.11.2015).

- **[Wells 2009]**: Website "Extreme Programming". URL http://www.extremeprogramming.org/donwells.html (26.11.2015).

About the Author

Stefan Luckhaus is a computer scientist with more than 35 years of experience. He works in software development since 1981 and graduated in Frankfurt, 1988, with the title of Dipl.-Ing. (FH). Subsequently, he was a freelancer for 10 years. Since 1998, Stefan Luckhaus is an employee of PASS Consulting Group. While working as a developer initially, he later managed development projects leading him to the USA, Singapore, India and various European countries. Today, Stefan Luckhaus is responsible for the competence center Project Governance, providing process engineering for the software development of the entire PASS group and conducting productivity and quality measurements - internally for more than 20 IT shops as well as on customer order. He is a member of the PASS group's R&D unit and has the status of a principal innovation consultant.

Stefan Luckhaus' fields of expertise are software metrics, quality management and process models / engineering for software development. In the German ICT industry association BITKOM (Bundesverband Informationswirtschaft, Telekommunikation und neue Medien e.V.) he chairs the work group quality management, collaborated in the publication „Agile Software Engineering Made in Germany" and has been a speaker, among others at the Bitkom Software Summit.

Stefan Luckhaus is present on the social networks LinkedIn, Xing and Twitter, is authoring the blog www.software-productivity.com (English) and co-authoring the blog www.it-management-blog.de (German).

www.ingramcontent.com/pod-product-compliance
Lightning Source LLC
Chambersburg PA
CBHW071216240526
45470CB00018B/2039